Immortal Again

Secrets of the Ancients
Their Secrets can Increase our Longevity

By
Walter Parks

UnKnownTruths
Publishing Company

Copyright © 2011
UnKnownTruths
Publishing Company

**UnKnownTruths
Publishing Company
8815 Conroy Windermere Rd. Suite 190
Orlando Florida 32835**

Library of Congress Control Number
(LCCN) 2004104317

Immortal Again

By

Walter Parks

ISBN-13:
978-1460981115

ISBN-10:
1460981111

"It is our duty, my young friends, to resist old age;

To compensate for its defects by a watchful care;

To fight against it as we would fight against disease;

To adopt a regimen of health; to practice moderate exercise;

And to take just enough food and drink to restore our strength

and not to overburden it.

Nor, indeed are we to give our attention solely to the body;

Much greater care is due to the mind and soul;

For they, too, like lamps grow dim with time,

unless we keep them supplied with oil."

Cicero
106-43 BC

****

"The overall deterioration of the body that comes with growing old is not inevitable."

Dr. Rudman
After HRT Tests
June 1989

Contents

Dedication

This book is dedicated to Betty. May you live forever.

Walter Parks

Preface

Much of the ancient literature of myths and legends relating to our many religions are held to be true by very few "believers". Some believe only a part of the myths that relate to **their** religion, but almost none that relate to other religions.

So the vast majority of "myths" are totally discounted by the vast majority of people.

Too bad.

Methuselah Depicted Near Age 969.

Most myths have origins in truths – great, **Unknown Truths.**

Did Methuselah really live to be 969 years old as stated in the Bible?

Yes, I think that he did. I believe that I've found the ancient secret of how he did it.

I also believe that we may soon be able to use his secret to increase our longevity.

How?

Please read, learn, and enjoy.

Walter Parks

Chapter I

Ancient Documentation

".... And all the days of Methuselah were nine hundred sixty and nine years ...."

Genesis 5:27

And the Bible tells us that others at that time also lived to such old ages, as shown in the table.

Patriarch Ages in the Bible

Patriarch	Age in Years
Adam	930
Seth	912
Enos	905
Cainan	910
Mahalaleel	895
Jared	962
Enoch	365
Methuselah	969
Lamech	777
Noah	959

Many of the old Biblical Patriarchs lived very long lives.

The clay tablets from ancient Sumer and Babylon say their Kings lived comparable life times. The tablets list their reigns rather than their ages, but we see from the chart that their life times were comparable to the Biblical patriarchs.

Reigns of the Sumerian Kings

King	Reign in Years
Gaur	1200
Gulla-Nidaba-Annapad	960
Pala-Kinatim	900
Nangishlishma	Tablet Damaged
Bahina	Tablet Damaged
Buanum	840
Kalibum	960
Galumum	840
Zukakip	900
Atab	600
Mashda	840
Arurum	720
Etana	Unknown

The Sumerian Kings are also said to have lived long lives.

Some of us were taught from childhood to believe in the Bible, based on faith alone. But others smile at the absurdity of living to be 969 years and attribute such statements to just another old myth.

But what is a myth?

> **"A story that is usually of unknown origin and at least partially traditional, that ostensibly relates historical events usually of such character as to serve to explain some practice, belief, institution, or natural phenomenon, and that is especially associated with religious rites and beliefs."**
>
> **Webster's**

How did this "myth" of great longevities get started? Did it have even a grain of truth at the beginning? Certainly no one lives to be that old today. But some argue that may be because the Bible told us that after the flood, no one would live past 120 years. God, speaking of man says:

> **".... His days shall be a hundred and twenty years."**
>
> **Genesis 6:3**

So no one seriously explores the question of Methuselah's age of 969 years. The believers believe on faith alone. But almost no one cares because it does not matter, since we cannot do it now, after the flood.

The doubters and non-believers dismiss it as not worth further consideration. But evidence that it may have been true keeps popping up. Re-reading the ancient literature and reviewing new archeological finds, gives new clues of authentication. And new scientific understandings suggest that these old stories may have some, actually a lot, of truth in them.

What could be the basis of such longevity?

Soma.

Soma was the (intoxicating) juice of the now unknown soma plant. In the Vedic lore of ancient India, it is said that:

> **" In the land of the fathers.... In the shining Paradise" that the Kings sat beneath a tree and drank soma while listening to flutes being played.**

They drank soma because it gave them immortality.

The "Shining Paradise" is believed to refer to the land from which the early Vedic peoples migrated after the great flood. The Shinning Paradise was a great civilization that existed prior to the flood. Plato said it was Atlantis.

Atlantis is described in the book and eBook:

Atlantis
The Eyewitnesses

The flood is described in the book:

Noah's Flood
When Where Why
The Conclusive Evidence

Atlantis or not – it was a civilization that existed prior to, and was destroyed by, the flood.

The ancient Rigveda of India devotes its entire ninth book to the praises of soma. It says that soma juice was the nectar of the gods; the elixir that gave them immortality. But it also says that mortals could use soma juice.

Soma was so important, that in the later legends of ancient India, it was elevated to a god:

> *This Soma is a god; he cures*
> *The sharpest ills that man endures.*
> *He heals the sick, the sad he cheers,*
> *He nerves the weak, dispels their fears.*
> *The faint with martial ardor fires,*
> *With lofty thought the bard inspires.*
> *The soul from earth to heaven he lifts,*
> *So great and wondrous are his gifts.*
> *Men feel the god within their veins,*
> *And cry in loud exulting strains.*
> *We've quaffed the Soma bright*
> *And are immortal grown.*
> *We've entered into light*

———

And all the gods have known,
What mortal now can harm,
Or foeman vex us more?
Through thee beyond alarm,
Immortal god, we soar.

Note especially:

"We've quaffed the Soma bright
And are immortal grown."

We drank Soma and became immortal! Is this just a wish? Could it be documentation of what they did?

Ancient Sumer and Babylon

We have also found references to this Soma plant in the Epic of Gilgamesh, written in the clay tablets of ancient Sumer and Babylon. These tablets note that they were copied from even more ancient texts.

They tell of King Gilgamesh who goes on a mission to seek eternal life. He is told of a plant that can provide him immortality. The ancient document refers to the Soma plant used by the "Kings and Nobles" prior to the flood.

In the ancient clay tablets re-write of several thousand years ago, the story says that the plant is now in deep water, probably an allegory for the flood, which destroyed the plant; made it become extinct.

Gilgamesh, being the hero of the story, of course dives and gets the plant, but it is taken away by the snake. This is the continuation of the allegory in that the snake caused the food and destroyed the life prolonging Soma plant. The snake and the flood are described in detail in the book about Atlantis referenced above.

The Bible says that – before the flood – such a plant grew in the Garden of Eden:

> *"And the Lord God planted a garden eastward in Eden...."*
>
> *Genesis 2:8*

> *And out of the ground made the Lord God to grow every tree that is pleasant to the sight, and good for food; the <u>tree of life</u> also in the midst of the garden, and the tree of knowledge of good and evil."*
>
> *Genesis 2:9*

And we've been taught that that same old serpent induced Eve to eat of the tree of knowledge. And thus God kicked Adam and Eve from Eden:

> *".... Lest he put forth his hand, and take also of the tree of life, and eat, and live forever."*
>
> *Genesis 3:22*

So we see many references to this plant, this tree of life, in literature from almost all of the ancient cultures.

What does all this really mean? Is there any scientific evidence for any of this?

Yes. Let's take it a step at a time. Let's understand why we age and what we can do to become **Immortal Again, like the ancients**.

Chapter 2
Why We Age

Life is a moderately good play with a badly written third act.

Truman Capote

Much of this information on Why We Age is taken, with permission, from the book (also in eBook format),

"Aging is a Treatable Disease."

To live young forever has been man's quest throughout history. From among the first writings preserved in the clay tablets of Babylon we read of King Gilgamesh who searched for the secret of eternal life. The tablets tell us that he found it, but lost it to the snake – and died.

King Gilgamesh depicted in ancient art.

Ponce de Leon searched for the fountain of youth – and never found it.

But we are now beginning to understand aging.

Almost all life on earth blossoms with youth until it has reproduced and passed its genes on to the next generation. After that the flowers wilt and die, all animals age and die, and we humans began to age and die.

We age because the products of our metabolism, i.e., the "ashes" from the oxidation processes that produce energy in our cells accumulate faster then our aging endocrine system can remove them. This is because most of the cleansing hormones that surged through our youthful bodies begin to decrease as we begin to age. Some of these more critical hormones have decreased by about 10 to 30 percent as we enter our 30's.

The decreases become ever more dramatic as we enter successive decades of life. Most of our hormones have decreased by over 50% and some have been reduced to near zero as we enter our 70's. **So we age.**

Our muscles and bones weaken; our reaction time slows; we lose our agility; all combine to make us more susceptible to accidents. Our immune system weakens and makes us more susceptible to disease. **And we die.**

There are a dozen or so mechanisms of aging that have been theorized over time. The author believes that there are seven basic causes that combine to make us age:

1. Free radicals and other ashes of our metabolism, and environmental toxins build up in our cells and cause the cells to die without being able to reproduce themselves as they would normally do when you are younger.

2. Our endocrine system ceases to secrete sufficient quantities of certain enzymes and hormones to keep up with the cell's battles with the build up of contaminants.

—

3. Our cells lose their ability to divide, and replace themselves because they use up their allotted number of divisions (reach their Hayflick Limits as explained later).

4. Stress causes secretion of excessive cortisol which does significant damage.

5. Some of us have inherited flawed genes that cause, or allow malfunctions.

6. Deficiencies in our diet limit the materials necessary for the cells to cleanse and repair themselves.   Excesses in our diets also adversely affect certain chemical reactions.

7. Lack of exercise causes atrophy of critical muscles that result in chemical imbalances and loss of strength and agility which makes us prone to accidents.

All animals studied to date share these same causes of aging that they and we have inherited from our common evolution.   Let's look at a little more detail for each of these 7 causes of aging.

1. Free Radicals, Lipofuscins, and Toxins

Mitochondria, located in almost all of your cells, convert the food you eat into energy. These products of your metabolism, i.e. this conversion to energy, also include "ashes" from the oxidation process that produces energy in our cells.

Some of these ashes are lipofuscins and free radicals that accumulate in our cells. These free radical molecules cause dangerous inflammation in the cell. Lipofuscins are the brown age spots we see on our skin. Similar "spots" and contaminates accumulate in our heart muscles, brains and elsewhere.

Most of the free radicals are a form of oxygen that "wants" to keep on "burning" with something. They attack various parts of the cell that they are in, including the cell walls and even the DNA. This eventually will cause the cell to become inoperative and die.

This oxidation is analogous to rusting. It results in a "rusting" of your arteries, which is partly responsible for the aging of your cardiovascular system.

Further, many various types of environmental toxins find their way into our cells and frequently behave as free radicals and/or accumulate to the point of clogging the cell and impeding its operation. Some of the toxins that we encounter are potentially very harmful and can cause cancer, asthma, various allergies and can significantly reduce your quality of life. Their affects tend to speed up the aging process.

Our bodies naturally combat these processes by providing anti-oxidants from the food we eat and supplements that we may take. Such anti-oxidants include the well-known vitamins C and E that we get by eating certain fruits and vegetables and by supplementation.

And there are many more free radical fighters in the form of many dozens of other anti-oxidants and nutrients. The best fighters are the enzymes and hormones described below.

But we lose our best fighters as we age, and the free radicals eventually win their battles. We age and die.

2. Hormone and Enzyme Decline

Our hormones regulate and control most of the functions of our bodies. Testosterone and estrogen, the major sex hormones in men and women respectively, give us the urge and ability to reproduce and continue the survival of our species.

But once we're past our reproductive prime, our hormone levels drop. This results in a lack of sex drive, in insomnia, impotence, weight gain, and countless other potential health problems that significantly decreases our quality of life.

So we see that our hormone system was designed primarily for reproduction for the survival of the species. Our bodies produce high quantities of certain hormones and enzymes during our youth. These give us our youthful vitality, strength, and endurance. They help in the battles against free radicals and they help provide nutrients for cell repair. They keep our cells cleansed of the ashes of metabolism and environmental toxins.

As long as our bodies produce sufficient quantities of these enzymes and hormones, we stay young. But we, and all plants and animals were designed to stay healthy until we have reproduced and reared our young. Mother Nature has little interest in us after we have passed our genes on to the next generation and cared for our children until they can fend for themselves.

As we age past our prime reproductive years we are no longer capable of producing sufficient quantities of the enzymes and hormones required to keep our cells "young and fit." With too little of these substances, our cells begin to lose their battles against the free radicals and other destructive elements.

The cells begin to age, and die. The organs of which they are a part become ineffective. We become frail, we die.

3. Telomere Erosion

As our cells age and become clogged with the ashes of metabolism and environmental toxins, they must replace themselves as they begin to become inoperative and begin to die. This is where telomeres come into play.

Our chromosomes which carry our DNA are in the center nucleus of our cells. At each end of each of the paired chromosomes there is a telomere. It is analogous to the grommets at the ends of shoelaces. The telomeres keep the chromosomes organized.

As the cells age and become clogged the cell divides to reproduce its self. In so doing, it loses some of the telomere length.

Dr. Hayflick determined that most normal cells, after the initial rapid growth of the embryo, can divide on average about 50 more times after birth before their telomeres are too short for further divisions.

As our cells age and become less effective, they divide to produce new cells to replace themselves. But the more times that they divide, the less effective the cells become at reproducing themselves. Eventually their telomeres become too short for the cells to reproduce.

Baring poor nutrition and a poor environment, the telomeres of most of our cells may allow replacement for up to the age approaching 120 years.

Some have noted that this age limit is not unlike the limit of life span quoted in the Bible. This limit is now termed "The Hayflick Limit" because Dr. Hayflick was the one that discovered the telomere function.

Treatments that address the so-called normal aging mechanisms, and the standard treatments that provide age reversal, do not affect the Hayflick limit. This is a separate aging mechanism. Dr. Hayflick, and others, believes that we cannot live past 120 years of age. The author disagrees and will discuss in more detail in a later Chapter.

4. Stress

A study in 2004 first identified the direct link between stress and aging. Intense, long-term emotional strain and stress can make you get sick and grow old much faster than normal.

Stress is a normal part of life. Almost everything that happens to us puts stress on our body. You can experience stress from your environment, your actions or inactions, the actions or inaction of others, your body aches, and even just your thoughts.

The human body is designed to experience stress and to react to it. Stress has a positive side in that it keeps us alert and ready to avoid danger. Stress becomes negative when you face continuous challenges without relief or relaxation between challenges. As a result stress-related tension builds.

Our adrenalin system was designed to handle **acute stress** which is experienced in response to an immediately perceived threat, such as a lion about to eat us. During an acute stress response, the autonomic nervous system is activated and the body experiences increased levels of cortisol, adrenalin and other hormones that produce an increased heart rate, quickened breathing rate, and higher blood pressure. Blood is shunted to the big muscles to prepare the body to fight or run away.

This is known as the fight-or-flight response.

But our lives have changed. Now we are more likely to be subjected to **chronic stress**, which our bodies were not designed to handle.

Many of us experience chronic stresses caused by our modern lifestyle. It can be caused by everything from high-pressured jobs to loneliness and to traffic snarls. Such chronic stress can keep the body in **a state of perceived threat**. When this happens, our fight-or-flight response, which was designed to help us fight a few life-threatening situations spaced out over a long period, can wear down our bodies and cause us to become ill, both physically and emotionally.

Chronic stress, sometimes termed "distress", causes negative kinds of stress reactions. It occurs when stress continues without any relief. Distress can lead to physical symptoms including headaches, upset stomach, elevated blood pressure, chest pain, and problems sleeping.

Research shows that stress can also cause or worsen some diseases.

Stress causes the release of the so called "stress hormones" of which cortisol is the primary. Chronically elevated levels of cortisol **speeds up the shortening of the telomeres** and causes the cells to become ineffective in carrying out their purposes. And as previously described, we know that shortened telomeres speeds the body's deterioration and causes aging.

Cells with shortened telomeres cannot divide to form new cells so they die without replacing themselves. The tissue to which the cells are a part then ages much more quickly and soon dies. And then we die.

So stress prematurely shortens our telomeres and speeds up aging.

This helps explain the association between psychological stress and increased risk of physical disease. Unrelenting emotional pressure and stress definitely accelerates the aging process.

5. Defective Genes

Defective genes are one of the factors affecting how fast and how well we age.

The Human Genome, which is a complete copy of the entire set of human genes, was basically decoded in 2003. With this decoding we now know, at least for many genes, which ones cause, and/or allow, the major diseases of man. We can do a DNA profile analysis and detect an individual's inherited flaws. Such flaws or defective genes will indicate the diseases and/or the abnormalities that an individual may be susceptible to because of his or her genetic inheritance.

We have developed treatments to replace defective genes with some success. The first treatment occurred in 1999 and many more treatments have occurred since. All of these have been experimental and conducted by researchers. It will be a while longer before the procedures will be available to the general public.

6. Dietary Deficiencies

Most of us in today's society do not experience dietary deficiencies unless or until we get older. But how fast, and well, we age will be affected when and if we do experience such deficiencies.

Our diets consist of five nutritional groups:
1. Proteins
2. Carbohydrates
3. Fats
4. Vitamins and supplements
5. Minerals

There are about 50 nutritional items within these groups that are necessary for good health. Age, gender, and overall health dictates how much of each of these nutrients that you need.

We need to insure that we get all of the nutrients that we need from each of the 5 groups to prevent any deficiencies. Relatively simple blood tests can determine any deficiencies and can monitor for any developing deficiencies.

When deficiencies exist they can be treated by diet changes and with supplements.

The causes of aging are described in more detail in the book and eBook, **"Aging is a Treatable Disease".**

Chapter 3
Telomeres, our Death Clock

The idea is to die young as late as possible.
Ashley Montagu

The key to limited life is the telomeres. The key to greatly extended life is the telomeres.

We know that as our cells age and become clogged with the ashes of metabolism and environmental toxins, they must replace themselves as they begin to become inoperative and begin to die.

Our chromosomes are in the center nucleus of our cells. At each end of each of the paired chromosomes there is a telomere. It is analogous to the grommets at the ends of shoelaces. The telomeres keep the chromosomes organized.

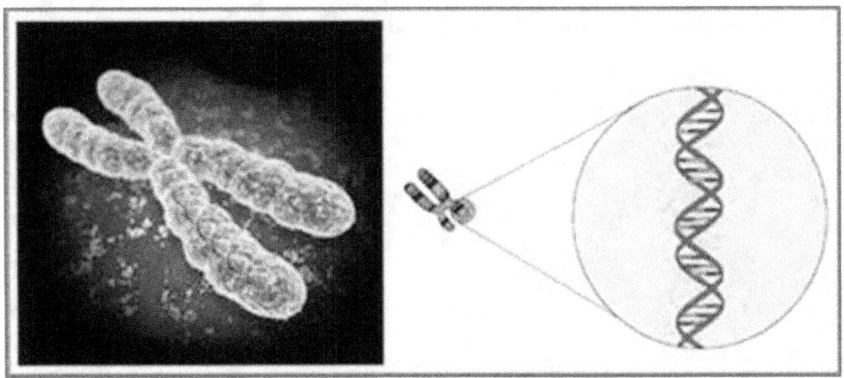

Chromosomes and our DNA

As the cells age and become clogged the cell divides to reproduce its self. In so doing, it loses some of the telomere length.

But the more times that they divide, the less effective the cells become at reproducing themselves. Eventually their telomeres become too short for the cells to reproduce.

This limit when the cells can no longer reproduce is now termed "The Hayflick Limit" because Dr. Hayflick was the one that discovered the telomere function.

It is interesting to note that Dolly, the famous cloned sheep died earlier than expected. That is because the cells from which Dolly was cloned were already several years old. They reached their Hayflick limit earlier than a naturally born sheep with young genes would have.

So what can we do to "reset" our telomere death clock?

Chapter 4
Resetting The Death Clock

Our hope of immortality does not come from any religions, but nearly all religions come from that hope"

Robert Green Ingersoll

King Gilgamesh, the ancients in Sumer and Babylon and Methuselah and the ancient Biblical Patriarchs, according to ancient literature, found the means to live very long lives.

Their ancient knowledge and the means to extend longevity were lost, most likely due to the great flood when almost all of mankind was destroyed.

We are just now beginning to learn what they knew in antiquity. The relatively recent decoding of the Human Genome is helping us to much better understand our genes and how they work.

It is very interesting to apply our new understandings to the ancient technology.

Understanding the double helix and telomeres of our DNA helps us understand what the ancients knew over 10,000 years ago.

Resetting the Death Clock

Tests over the past few years have indicated that the "Hayflick limit" may be extended by the use of an enzyme that causes the "organizing genes" at the ends of the chromosomes (the telomeres) to re-grow. This enzyme is called telomerase.

We learned about telomerase from one of our worst enemies, cancer.

A few years ago, cancer researchers noted that some cancer cells divide and reproduce extremely fast, and many more times than the Hayflick limit. It was learned that these cancer cells secrete an enzyme named telomerase. Telomerase causes the telomeres at the ends of the chromosomes to grow such that their cells can continue to divide.

These cancer cells appear immortal.

The researchers became both excited and concerned; excited that cells could become immortal; concerned that such cells would always become cancerous.

A series of experiments indicated that their concern was, perhaps, unfounded. The experiments consisted of treating various cells, including human cells from the foreskin of circumcised infants with telomerase. It was proved that telomerase could be used without inducing cancer.

These human cells were injected with telomerase and grew longer telomeres. There was no cancer. The cells continued to divide many times beyond their Hayflick limit. So far the cells have divided the quantity of times equivalent to a human life of over 1500 years.

These human cells appear immortal.

Telomerase therapy promises to extend life beyond the 120-year Hayflick limit.

Doctors and researchers involved in these treatments are reporting that it is their belief that eventually death will not be inevitable.

Someday, when sufficient effectiveness and safety tests have been completed, telomerase treatments are expected to become available.

Then what will we do?

The first and safest use of telomerase treatments will probably be to rejuvenate damaged skin.

Some areas of the body suffer more "stress" than do other parts, and therefore use up their Hayflick limit sooner. An example is the rough skin on the hands and necks of field workers that have experienced long exposures to the sun. The skin cells continuously reproduce – divide – to correct the continuing damage, and thus use up their Hayflick limits. The consequence is that the back of the neck and the tops of the hands appear to age faster than does the rest of the (un-exposed) skin.

Telomerase treatments are expected to become wide spread after the treatments on sun-damaged skin prove their effectiveness – and safety.

We are still some years away from being able to use telomerase treatments for more complex procedures. Extensive tests are required. Some experimenters, however, believe that telomerase treatment will eventually offer the promise of extending life beyond the 120 year Hayflick Limit. Some even believe that it will some day offer eternal life.

Many researchers believe that the first immortals are alive today.

Chapter 5
The Tree of Life

The average man, who does not know what to do with his life, wants another one, which will last forever.

Anatole France

So maybe there was a tree of life.

If so, it was most likely soma.

We see in the art of the ancients many depictions of the tree of life.

The Tree of Life

Here we see a god picking leaves from the tree of life. He is holding a pine cone which was the symbol for long life. He is also holding a "purse" in which, it is speculated, he carries his soma, tree of life, leaves.

The soma plant apparently became extinct when the great flood destroyed most life on earth in about 9619 BC as described in the books:

Noah's Flood
When Where Why
The Conclusive Evidence

Atlantis
The Eyewitnesses

Here we see a better view of a god holding his purse of soma.

Ancient god with purse of soma

And here are two more carvings from antiquity of gods holding their purses of soma.

The theme of gods, and kings, holding the pine cone as a symbol of long life, and with their soma pouches is repeated through antiquity.

And this one is my favorite.

Winged god with soma purse

Can we become **Immortal Again** like the Biblical Patriarchs and Sumerian Kings of yesterday were before Noah's Flood destroyed their Soma?

Can we combine the lore of yesterday with the new DNA and Human Genome knowledge of today to recreate **The Tree of Life** and become **Immortal Again**?

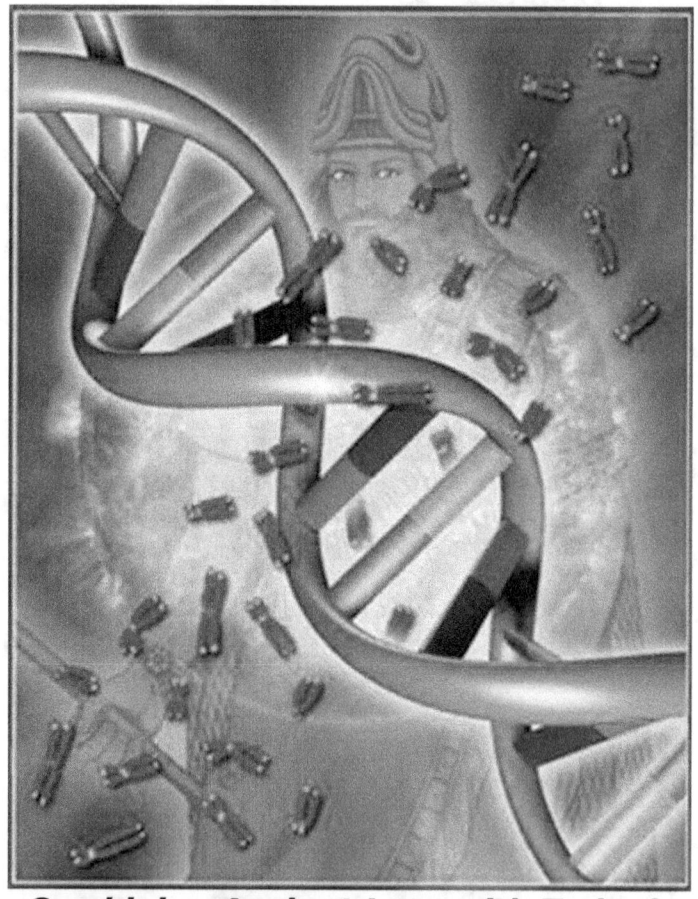

Combining Ancient Lore with Today's Knowledge of DNA may enable us to become Immortal Again.

Chapter 6
What We Know About Immortality

In these matters the only certainty is that nothing is certain.

Pliny

We know the following about immortality.

Dr. Hayflick determined a few years ago that the maximum "natural" age of man is about 120 years. This is the age that man can live to be if diseases or accidents do not get him earlier. This 120 year "Hayflick limit" is set by the length of a set of genes at the end of the chromosomes in man's DNA.

To stay healthy, our cells must be periodically replaced to provide new cells as oxidants, toxins, and normal wear and tear mechanisms age and destroy our cells.

When a new cell is needed, a healthy cell divides its chromosomes (DNA) and splits into two cells – thus providing a new cell. Each time the chromosomes split, they lose part of their genes at the ends of the chromosomes. These genes are necessary to keep the chromosome structure organized. When these end genes (telomeres) become too short, that is after they have divided a given number of times, the chromosome can no longer divide. No new cells can be produced. The existing cells age and die. Since they cannot be replaced, we age and die.

The Bible and other ancient literature tells us that the Biblical Patriarchs and Sumerian Kings lived very long lives before Noah's Flood, but was limited to 120 years after the flood.

Apparently the Biblical Patriarchs and Sumerian Kings got something from drinking Soma juice that made their telomeres continue to grow.

Our telomeres continue to grow while we are embryos in our mother's womb. They have to re-grow to allow for the rapid growth of the embryo while in the womb.

Embryos get telomerase in a natural manner so that they can sustain their rapid growth. Our bodies do not naturally produce telomerase after birth.

Dr. Harley, et al at the Geron Corporation recently used the telomerase enzyme in lab tests to re-grow eroded telomeres in human cells. The cells then divided (growing new cells) many more times than the natural Hayflick limit would normally allow. They made human cells become immortal by the use of telomerase.

The protein structure of telomerase is similar to an enzyme found in a certain plant. A "cousin" of this plant with the exact protein structure of telomerase, may have lived before - and was destroyed by - the flood.

So how did Methuselah live to 969 years?

It is speculated that Methuselah drank **Soma** juice from the **Tree of Life** and thereby ingested the **telomerase** enzyme from the plant.

The plant got wiped out, became extinct because of Noah's Flood.

Now after the flood no one can get the telomerase enzyme so our telomeres at the ends of the chromosomes continued to erode as the cells split to form new cells. After a certain number of splits the telomeres become too short to continue to divide and our life span is limited to our current "natural lifespan" of 120 years.

The Bible tells the same story, but in a more poetic manner.

Chapter 7
Immortal Again

I don't want to achieve immortality through my work. I want to achieve it through not dying.
Woody Allen

Can we become Immortal Again?

Yes, probably, eventually.

There are at least 3 potential approaches to becoming Immortal Again.

1. Genetically Produce a Soma (Telomerase Plant)

2. Find a way to re-activate, i.e. turn back on, the genes that enabled us to produce telomerase while embryos in the womb

3. Clone bacteria to produce human telomerase in the manner in which we have cloned bacteria to produce insulin for diabetics and for human growth hormone (HGH) for the anti-aging programs.

Genetically Produce a Soma (Telomerase Plant)

The protein structure of telomerase has been found to be very similar to an enzyme found in a certain plant. This research work is proprietary so the name of the plant is considered Company confidential.

A "cousin" of this plant with the exact protein structure of telomerase, is believed to have lived before - and was destroyed by - the flood.

It is proposed that we genetically modify this plant to produce the telomerase enzyme. We would then use one of the already developed methods to get the telomerase to all of our cells.

We would then run a ton of tests on animals and then humans to verify the effectiveness and safety.

This approach appears very feasible.

Conventional breeding techniques have been in practice for hundreds of years. These techniques are now giving way to genetic alterations. Genetically altered crops have been planted since 1995.

In 1955 a gene from a strawberry plant was inserted into a mustard plant so it would produce a chemical lure to get predator insects to come to it to protect it from mites and other insects harmful to the plant. This genetic modification delivers the plant's defense against insects in its seed rather than having the farmer have to use a spray. This is because the genetic modification becomes part of the plant's DNA; the modified plant becomes a new plant; its life is perpetuated in its seeds.

The study team reports that the genetically engineered plant was able to attract predatory mites (a small relative of spiders) that prey on plant-eating spider mites. The plant needed little genetic modification to introduce the chemical lure. The machinery to produce this "alien protein" is available in all plants. You just have to tap into the existing pathways [in plant cells], and that may be done by the introduction of just one gene.

This research is expected to allow humans to harness plant-bug interactions to improve essentially all future crop yields.

During the last five years, several large companies, including Monsanto and Syngenta have altered our food supply with genetic modifications. Genes from bacteria, viruses, foreign plants and animals have been inserted into corn, soybeans, potatoes, tomatoes, squash, and papayas. These corporations plan to "genetically engineer" almost 100% of our food within a decade.

Already about 40% of the soybeans, 20% of the corn, and a percentage of the potatoes grown in the U.S. and about 50% of the canola (rapeseed) plants in Canada have been genetically altered.

More than 60% of the packaged food items in US grocery stores contain genetically altered ingredients.

Scientists have known how to get plants to resist pests and herbicides for years. It was demonstrated in 1995. Genes from other plants and animals can be inserted to allow the altered plants to produce their own insecticides.

What is new in 2011 is the ability to induce plants to create new products by tinkering with the plants' own synthetic pathways.

Sarah O'Connor of the Department of Chemistry at MIT has studied periwinkle for several years because it produces a variety of compounds that may be used to treat cancers and hypertension.   She has now produced new compounds by manipulating the complex biosynthetic pathways of the periwinkle plant. This sort of manipulation offers a new way to tweak potential drugs to make them more effective and less toxic.

Earlier, in 2007, several plants were genetically modified to remove toxic compounds from the environment.

One of the research groups used small plants related to cabbage and mustard to clean up soil contaminated with cyclonite, or RDX. The widely used RDX explosive is highly toxic and carcinogenic.

Another research team modified a poplar tree to soak up a host of cancer-causing compounds from soil, groundwater, and air. The genetically altered plants break the contaminants down into harmless compounds in a process called phytoremediation.

"It is our hope that by developing trees that can remove carcinogens from the water and air in a fast and economical way, people will be more likely to use [the land] than abandon the property as too expensive to clean up," said Sharon Doty, of the University of Washington.

It should also be noted that we have also placed the genes from a pig into a tomato to provide the tomato with a longer in-store shelf life before it begins to soften and rot.

With all of this research and actual applications, especially in the food crop industry, it will just be a matter of time before the genetic alteration of plants to produce telomerase is accomplished.

It will necessarily be led by the development of an effective and safe method to insert the telomerase into the cells of our bodies.

Develop Method to Insert Telomerase in Our Cells

Various techniques have already been used to alter genes by inserting genetic materials from other plants and animals into the gene to be altered.

For plants, the most common vector is a bacterium called *Agrobacterium tumefacsiens.* The bacterium has the ability to incorporate DNA into plant genetic material. This is a possible way to inject the altered gene for the production of telomerase into the plant to be used to produce telomerase.

Viruses also are good gene vectors for both plants and animals. The viral infection process involves the integration of viral genetic material into the host genome. Viruses can thus be engineered to deliver DNA, which has been introduced into the viral genome prior to infection.

The altered virus then "infects" the plant with the genes to produce telomerase.

Gene insertion can also be achieved mechanically by microinjection of the gene into a **cell**, or ballistically, by shooting gold beads coated with the gene of interest into the cell. This is not controllable enough for the telomerase treatment in developed humans. It could be used at the embryo stage.

One or more of the more recent developments in gene therapy research will most likely be the methods first used to inject telomerase in developed humans.

Good results have already been experienced in a process called adoptive cell transfer. Researchers drew a small sample of blood that contained normal lymphocytes from individual patients and infected the cells with a retrovirus in the laboratory. The retrovirus acts like a homing pigeon to deliver genes that encode specific proteins, called T cell receptors, into cells. These make more receptor proteins to coat the outer surface of the lymphocytes. The T cell receptors act as homing devices in that they recognize and bind to certain molecules found on the surface of tumor cells. The T cell receptors then activate the lymphocytes to destroy the cancer cells.

In this test, the newly engineered lymphocytes were infused into 17 patients with advanced metastatic melanoma. One month after receiving gene therapy, all patients in the last two groups still had 9 percent to 56 percent of their T cell receptors-expressing lymphocytes. **There were no toxic side effects attributed to the genetically modified cells in any patient.**

In 2005 Gene Therapy successfully cured deafness in guinea pigs. Each animal had been deafened by destruction of the hair cells in the cochlea that translate sound vibrations into nerve signals. A gene, called Atoh1, which stimulates the hair cells' growth, was delivered to the cochlea by an adenovirus. The genes triggered re-growth of the hair cells and many of the animals regained up to 80% of their original hearing thresholds. This study, which many pave the way to human trials of the gene, is the first to show that gene therapy can repair deafness in animals.

In early 2003 a research team at the University of California, Los Angeles was able to get genes into the brain using liposomes coated in a polymer call polyethylene glycol (PEG). The transfer of genes into the brain is a significant achievement because viral vectors are too big to get across the "blood-brain barrier." This method has potential for treating Parkinson's disease.

Sickle Cell Disease was successfully treated in mice in 2002.

This research and clinical studies are rapidly producing a body of work that can lead to the successful injection of telomerase in perhaps a shorter time than previously thought

Re-activate Our Telomerase Genes

Another approach to getting telomerase as we get older is to re-activate the genes that produced telomerase when we are embryos in our mother's womb. All of us carry these genes when we are embryos but cease to produce telomerase after birth.

We need to find a way to re-active, i.e. turn these genes back on especially so as we get older.

The decoding of the human genome in 2003 provided us with the fundamental understanding of what our genes are and what they do. We need more studies to go from this fundamental understanding to a detailed understanding of the genes that produced telomerase when we were embryos.

And perhaps even more importantly we need to understand what turned these genes off and what we need to do to turn them back on.

This could be the best of all methods for extending our life spans and maybe even for achieving immortality.

We already know, at least to some extent, how genes are turned on and off. Turning genes on and off is a major activity of all living cells. Almost 10 percent of the genes in the human genome are switches for turning genes off or on.

The University of Michigan at Ann Arbor first explored the mechanisms for turning genes on and off 55 years ago. They studied gene regulation in bacteria and discovered that sugars in the food supply turn on the genes required for their own digestion. In addition, when bacteria are transferred from a medium containing the sugar lactose to a medium without lactose, the bacteria turn off their lactose-metabolizing genes.

The mechanism for this switch is based on a regulatory protein called the lac repressor.

In the 1970s and 1980s, scientists discovered that gene regulation in mammals also uses the mechanism of protein recognition of short DNA sequences. These short regulatory sequences are called enhancers. For example, the hormone testosterone binds a receptor protein that recognizes a 15-base-pair DNA sequence. As a result, genes that contain this sequence can be activated by testosterone.

Estrogen, in contrast, regulates a different set of genes that have their own distinct sequence. Researchers can exploit enhancers in experiments by fusing a known enhancer to a gene that they want to regulate.

As an example, one might fuse the estrogen enhancer to the hemoglobin gene and insert the construct into the chromosome of a mouse. When the resulting mouse is treated with estrogen, the hemoglobin gene will be turned on.

Some regulatory proteins control embryonic development. These are the genes we need to study to find a way to re-activate the telomerase gene.

Other regulators respond to changing environmental signals, such as the amount of protein or trace metals in the diet. Scientists are using the knowledge about gene regulation combined with some newly developed cloning tools to investigate the physiological effects of switching specific genes on or off in aging adults.

A California biotech company has a technology that can turn any gene on or off. The technology allows a scientist to genetically engineer a protein with what is called a zinc finger. Heart not pumping hard enough for lack of good blood vessels? Turn on the blood vessel-growing gene. Want to stop a patient from getting fatter? Turn off the gene that makes fat cells.

But biology is not yet up to speed with zinc finger technology. But it is comforting to believe that someday I will be able to re-grow arteries to replace the ones I have clogged with my undisciplined lifestyle; and to finally get rid of my excess weight.

Zinc fingers occur naturally inside the nucleus of all organisms, where they bind to DNA to turn genes on or off. They are the most common vehicle that genes use for alternating what protein they do or do not produce.

It is just a matter of time before all this new technology and findings are used to develop a telomerase treatment.

It is just a matter of time before we can become **Immortal Again.**

Time Magazine boldly stated on their cover of February 21, 2011:

2045:
The Year Man Becomes Immortal.

Can you last that long?

Maybe you had better start some of the more readily available anti-aging programs to insure that you will be around when the immortal treatments become available.

A companion book (also in eBook format) **Aging is a Treatable Disease** lists several program options to help you survive until then.

1. Understand and put into practice a basic, proven anti-aging program specifically tailored to you. **The future of medicine is in personal tailoring.**

2. Understand the severity of stress and **Practice Stress Therapy.**

3. Understand and consider personally tailored **Hormone Replacement Therapy (HRT)** which has proven very effective, but that **may be risky**.

4. Understand and consider personally tailored **Hormone Precursor Therapy (HPT)** which has proven effective, but less so than HRT. A few also believe it may be somewhat risky since it has not had a long history.

---

5. Understand and consider the **Calorie Restriction Program** which can be effective but is far too uncomfortable for almost all people wishing to lead a normal life.

6. Understand and consider applying for **Telomerase Therapy** when is becomes available.
Understand and consider, if it applies to you, **Gene Replacement Therapy.**

7. Understand and participate in one or more of the as yet **unproven "beta" programs** resulting from the newly evolving technologies, when they become available.

8. Develop your own **Personally Tailored Program** that includes your selections from various combinations of all of the above. This will be your **"full life plan"**.

Take a look at the book or eBook to see your options.

About the Author

After he retired from a career as an aerospace engineer and an aerospace executive, Mr. Parks decided to study two areas that he found very interesting. He continued his lifelong hobby of studying the available information to learn as much as possible about the knowledge of the ancients.

He also became interested in today's emerging anti-aging technologies.

He began to learn new truths about our modern day discoveries in science and social cultures that have amazing parallels with knowledge of the ancients. He found truths of the ancients that had become unknown down through the ages. Some of these "rediscoveries" included ancient practices that resulted in great longevity for some ancient leaders that had access to these practices.

He further learned that major advances in anti-aging are continuing to be made almost daily. But few people are aware of these advances and even fewer are taking advantage of the new findings

Rediscovering these ancient Unknown Truths inspired the formation of **UnKnownTruths Publishing Company** and the writing of several books on the subjects of ancient knowledge, philosophy and religion.

Mr. Parks also felt that the new discoveries in anti-aging, in a similar manner, represented Unknown Truths that prompted him to write books on these new discoveries. **Immortal Again** is one of these books and it also combines some secrets of the ancients.

Basic Education
BS Aeronautical Engineering, Mississippi State University, Starkville, Mississippi

MBA Business Administration, Rollins College, Winter Park, Florida

Post Graduate Studies:
Astrophysics, UCLA, Los Angeles, California

Laser Physics, University of Michigan, Ann Arbor, Michigan

Computer Science, University of Florida, Coral Gables, Florida

Finance and Accounting, The Wharton School, University of Pennsylvania.

Informal Education:
Mr. Parks has had a life-long hobby of studying our ancient pre-history.

Mr. Parks has engaged in an intensive study of the human aging process during the last ten years. This has included detailed studies of many of the scientific tests that have been conducted relating to cellular biology, the Human Genome, and Stem Cell Research.

Career Experience

Mr. Parks begin working for Martin Marietta Aerospace Company, now Lockheed Martin Aerospace Company, immediately after graduating from college. He progressed from an aerospace engineer to Vice President of the Company and President of the Tactical Weapons Division. He left Lockheed Martin after 24 years to form Parks Jaggers Aerospace Company and sold it 4 years later. He then formed Quest Studios, Quest Entertainment and Rosebud Entertainment to produce films at Universal Studios. He then formed UnKnownTruths Publishing Company to publish books of the previously unknown.

1958-1982 Aerospace Engineer and Executive at Lockheed Martin, with the last six years as Division General Manager, Tactical Weapons Systems.

1982-1986 Co-founder and Chairman of Parks Jaggers Aerospace Company which he sold in 1986.

1986-1996 Producer of 10 films, Director of 7 films, and Writer of 5 films produced at Universal Studios.

1997-2002 Writer and Researcher of ancient studies and anti-aging studies.

2003-Present Founder and President UnKnownTruths Publishing Company. Writer of 4 published books and 4 published eBooks, Researcher.

About
UnKnownTruths
Publishing Company

UnKnownTruths Publishing Company was formed to publish true stories of the unusual or of the previously Unknown or unexplained. These stories typically provide radically different views from those that have shaped the understandings of our natural world, our religions, our science, our history, and even the foundations of our civilizations.

The Company's stories also include stories of the very important anti-aging, life-extending medical breakthroughs; stem cell therapies; genetic therapies; cloning and other emerging findings that promise to change the very meaning of life.

The Company also publishes stories from the past that are so unbelievable that they are generally considered to be myths. The published stories provide the evidence for the truth.

The Company currently has an additional 9 books and eBooks in development.

www.ingramcontent.com/pod-product-compliance
Lightning Source LLC
Chambersburg PA
CBHW060224290526
45789CB00003B/1400